Amazon Echo
Ultimate Amazon Echo Beginner Guide

Simon Monty

Simon Monty

paraphrase any part or the content within this book without the consent of the author or copyright owner. Legal action will be pursued if this is breached.

Table of Contents

Introduction 7
Chapter 1 What Is the Amazon Echo? 9
Chapter 2 Benefits of the Amazon Echo 15
Chapter 3 Practical Applications 29
Chapter 4 Using the Amazon Echo 47
Chapter 5 Tips and Tricks.. 77
Chapter 6 Frequently Asked Questions 91
Conclusion.. 97

Simon Monty

Introduction

It was not very long ago that a person could not imagine having a phone to carry in their own pocket. It was only about two decades ago that the idea of the cell phone came into the world. The moment it did, no one imagined being able to have it run as fast as a computer or connect it to the Internet. Today, the cell phone has updated itself beyond most levels as people can carry all the information about their lives right in their purse or pocket.

A few years ago, being able to have technology speak to you and then go through to control the home was not possible to truly think about. It was an idea that was difficult to grasp. Now there are laptops, tablets, and phones that can connect documents together through the cloud within seconds. The same concept of connecting devices together makes the home another form of a device.

Today, the Amazon Echo is the same as a tablet that can connect a home to a hands-free speech technology system. It is a way to have a voice command system to make life easier. It allows for a person to multi-task at new levels and they are now able to be installed by any person. Even if a person cannot figure out a smart phone, this piece of technology with allow for them to install the device in their own home and they will then control more aspects of their surroundings. From remembering to lock the doors, to turning the lights on, to setting reminders, and then documenting groceries, the Amazon Echo is a personal assistant for your own home.

In this book, you can read about all the benefits and a step-by-step guide to the Amazon Echo. Each chapter will provide enough information to give the reader enough knowledge to buy their own device and make their life one step easier.

Thank you for purchasing this book and I hope you enjoy reading about all the features that the Amazon Echo has to offer.

Chapter 1
What Is the Amazon Echo?

Because of the many types of technology that have been created over the past few years, many different devices are now available. A voice recognition system in your home can be from a smartphone, a Google device, or an Amazon device. Here, it is highly recommended to buy the Amazon devices, since they are known to have the most voice recognition software that allows a person to speak from any distance without having to completely raise their voice. Not only that, but the Amazon devices also have more popular specs and they are seen as higher rated works of technology.

There is much to learn about what each of the devices has to offer. To start, this device is the newest 2017 Amazon Echo. In previous models, they were able to have voice recognition software that could still perform to the highest standards. This device stands alone in its own category. It is a device that is a combination of a speaker system and a voice recognition system.

When you get the product in the mail and open up the box, it will look like nothing more than a simple speaker. As you go through and learn how to use the device, you will realize that it is much more than just a speaker. It is comes installed with 'Alexa' software that can guide you through the process of connecting your home to the speaker. It is about 9.25 inches tall and has seven microphones built into its system. Inside this device, there is also a 2-inch tweeter that can communicate back to the user, along with a 2.5-inch subwoofer to make the quality of the speaker even higher. With these two things, it will generate an even better sound as well as creating a more pleasant environment for the household. To finish things off, this whole device uses far field technology. Basically, the Amazon Echo uses this form of technology to register your voice from a longer range. For example, a Google device might only register your voice from 15 feet away while this device can register your voice from 30 feet away. Not only that, but it is capable of hearing your voice over other noises in the room. There is no more worry about whether the speaker can hear you over the loud washing machine or the television. Amazon was able to create a simpler device that can suit you in your own home.

The 'Alexa' system built into the Amazon Echo makes it a more intelligent device. The software makes it more personal when speaking, while the voice sounds more realistic. The software makes it so that the device is always on and always functioning. Even though this might sound as if it will always be listening and that it might always be trying to do something or always be trying to speak, this is not the case. Instead, using the word 'Alexa' is the wake word for the device. The Amazon Echo will not be waiting to hear any voice to turn on. It will wait to hear the word 'Alexa' and then the next given command. This alone will ensure more privacy for the user to ensure that the device will not be able to pick up everything it hears. Since this is a time where people are fearful about how much technology is truly controlling their lives or how much their technology can track their words, pictures, or location, Amazon thought it through and created this device in a different way. While it may have many parts to the speakers and the sound system, it has to be opened through a word to ensure that the user is truly in the same room. It is the same as using a password to open up a laptop. Without the password, the user will not be able to get into their own computer.

Since there is so much to this device, it has to connect to the house through the processor. Doing this is as simple as connecting it to the Internet Wi-Fi. By using cloud-based processing to connect to the house, it can function faster with more efficiency. Through the Internet connection, this device can begin working. The voice of Alexa is created through a multitude of natural language processing (NLP) and algorithms through a text-to-speech program to automatically format to any voice that can be heard from the device. These two things together make it easier for the Amazon Echo to understand the user's communications and commands.

The device is one of the more efficient ways to have the house run and it can get more done for the user. Alexa, through the software programming, offers more than simply a smartphone or a computer. A smartphone can give the user more speed while browsing the Internet or simply communicating with other people. The same idea goes for a computer, where the use can open it and start working immediately. The Alexa software does not require an 'on' button and it does not require any sort of patience in opening apps because of a slower Internet or low data speed. Alexa can read the current news of the day, play your own favorite playlist of songs, or even give you updates on your favorite sport teams. There is no waiting around for the Internet or waiting for something to load. Instead, it is about saying the word 'Alexa' and then moving forward. This idea alone makes the Amazon Echo a faster piece of technology that can wire through the whole house to give the user a completely different experience from the phone or the computer.

Imagine life in the morning when, instead of leaving the kitchen while cooking to turn on the television, you can ask 'Alexa, what is the weather today?' or 'Alexa, what is the traffic life this morning?' Saving you the time in getting up to move from room to room, this device allows for the user to focus more on the task of that moment. No more will you have to worry about leaving the stove to look at the update of your favorite team. No more will you need to get away from reading your favorite book to find out what is next on your schedule for the week. No more will you have to worry about the accuracy of the different news channels to hear about what the traffic will be like that morning. The Amazon Echo creates a different life for you to have and it is capable of allowing you to put more focus on the current tasks at hand rather than running back and forth to see what the newest tasks of the day are.

With the integration of smart things, Phillips Hue, Wink, and Belkin WeMo, the voice control system will continuously be able to update itself with the best parts of the device. There will be no concern about when the device needs to be looked at or when it needs to be updated. Amazon can create the most current device to simplify any person's life. Whether they are trying to relax more or simply move forward to have another piece of technology within their home, both sides should look into the benefits of owning the Amazon Echo.

Chapter 2
Benefits of the Amazon Echo

With all the different voice recognition devices that are available, the question comes down to why a person should choose the Amazon Echo. Since there are many options that can be purchased, consumers have to think about all the different aspects of a device. They will sit down at the computer to find all the needed information before they make a final decision. Looking past the price points of any of the objects, they will see what device is best for them to handle. From simplicity, to upgraded software, to personal touch, and to what makes it unique, every part is considered before a person presses the button to 'buy now.' With all of this in mind, why should the Amazon Echo stand out above all the others?

At start, the first product ever introduced by Amazon since the Kindle was called 'The Echo.' Amazon did not intend to simply create an e-commerce business where they make the product to take the money from all the buyers right in front of them. Instead, they are looking to improve the market and to provide more options for all the merchants surrounding their own business. By diversifying the market, they can provide more of an outlook on what devices are better to own and they can give more accurate feedback for the buyer. Amazon will tell the truth about all of their devices to their buyers to ensure that they are completely satisfied. If not, they risk losing more customers instead of gaining a larger market share. One or two bad reviews will lead to less popularity, but it will allow for them to recover through a newer device. If they decided to use only fluff to describe their objects, then they would lose the trust of all their consumers and they would risk permanently losing customers.

The Echo itself was created over a span of a few months. Since there were many people looking into the upgrade of the device and looking into how to provide more satisfaction towards their own customers, they were more concerned about the customer rather than the product itself. They still continue to hold to their word of working to make the life of the consumer easier with a multitude of optimum services and easier forms of functionality for the buyer to experience.

When you consider buying the Amazon Echo, it is best to see how Amazon functions. They will work to provide the highest quality of service for each of their products. With a fast-growing company, they all work to have the greatest form of products that people can experience. If you have to go through a busy schedule of looking up when the next meeting is, then the Amazon Echo is the best device for you to buy. If you are simply curious about many things and are constantly looking on your phone or laptop for answers, then the Amazon Echo would be able to save you time. If you are looking for a speaker system that will act on your command to play your favorite music playlists, then the Amazon Echo is the right product for you. Being able to play your music, answer any of your questions, or say what is next on your calendar are only a few things that this device can do. The more you read through this chapter, the more you can see if the Amazon Echo is the right device for you to choose in your own home.

The first thing to realize when you consider buying this device is the ease of access that it provides. Once you open the box, there are very simple and easy-to-follow instructions inside. That means you will not have to look up online the tools or phrases needed to begin working with the device. Instead of focusing on what is in this book or what you can find online, you can begin using the device immediately. If you run into any kind of trouble, Amazon has set up a personal support team for this device alone. They are available to any time of day to ensure that your device will work perfectly. That means that it does not matter what time zone you are in or what your schedule, they can provide service to you at any point of the day. From there, following the instructions is as simple as choosing your personal preferences for the device and the settings that are best able to suit your own personal standards. From here, it is about turning on the device, connecting it to Wi-Fi, and listening to tell 'Alexa' what settings are best to have. The Echo will then be waiting for you to say either 'Alexa' or 'Amazon.' The command following either of these words will tell the device to handle your queries and then it will also assist you in any task that you have.

The second best part about the Amazon Echo is the voice itself. Being able to have an excellent voice system along with superior voice recognition provides a more stable outlook and a more efficient way for the user to get through the settings and daily tasks. The voice quality in this little device is more brilliant than the other competitors. With the subwoofer and the many speakers, this device is more powerful and gives more than the others. It can be your own personal music therapist while playing all of your favorite songs in excellent quality. The voice of 'Alexa' is crisp because of the amount of technology that went into this device. The moment you hear the voice, you will hear a crisp, loud, and soothing pitch that provides a better representation of what needs to be heard. Would you rather hear a robotic voice full of static, or a more personal voice that can bring a smile to your face every time you hear it? After having a long and tiring day of work, saying 'Alexa, play my music track' will brighten your day. Knowing that you didn't have to do anything more than saying one phrase to listen to your favorite music will begin to relax your mind. Instead of adding to a long list of things to do during the day, this device narrows it down to a simple phrase.

Another part of the voice is the superior voice recognition system built into the of 'Alexa' program. One of the common questions people will ask is: What makes the software and voice of 'Alexa' different from any other human voice? What makes this voice any different compared to the other devices from Google? Is she truly able to pick up your voice better than all the other devices out there? There are many questions that come from this one voice. Alexa was created to outrank all the other voice recognition software programs available. She can hear your voice from farther away. Even if the user has music or the television on, Alexa will still register that the voice is speaking and will listen. Not only that, but she can pick out your voice from a crowd of people. Not any person can simply say her name. Instead, the software was created only to respond to the voice of the user. Alexa is basically a more intelligent kind of personal assistant to have inside your home. She can record any phrase or command that you say and then she will respond in whatever manner is needed. The device is not built to have the constant need of beeping or talking. Instead, it was created to keep things simple by saying the fewest phrases possible. It will not have the need to constantly remind the user of notifications for their schedule. By the end of all the instructions and setting up the device, it will continue to remain silent until the user says the name 'Alexa.'

The third thing to think about with this device is privacy concerns. Since every person lives with the concern of being tracked down through their phone or having parts of it being recorded, the same fear occurs with this device. It is a voice recognition system, and therefore it does contain your voice. The Amazon Echo is very sensitive toward any phrases that are heard through the microphone. With this idea in mind, Amazon then created the keyword 'Alexa' and 'Amazon' to ensure that nothing private would actually be accidentally recorded. Unless the device has recognized those words, there is no concern about whether or not there was any information recorded. Amazon technical experts created this software to open through a type of password system. By only recognizing the voice and the keyword of the user, it can be a most protective device for the user to have. Let's say you own the device and you are not sure about all the information that was recorded. This is no problem since you can log into your won user profile, go to the 'manage my device' tab in the profile, and listen to any of the recordings that have been made by the device. You can choose to delete them or leave them there. The workers and managers at Amazon took privacy very seriously when they designed this part of the device. This allows the user can choose to refresh the device to ensure that any words recorded by the device are all marked to be deleted.

The fourth thing to realize when thinking about buying this device is the software upgrades. Whenever a person buys a computer or a phone, the device will have to go through an upgrade at some point. It is not the kind of upgrade that requires the user to buy a completely new device. Instead, it is an upgrade that keeps the software up to date. In order to have all of technology connected and working properly, every piece of technology goes through either automatic upgrades or manual upgrades. With automatic upgrades, the user will not need to worry about if the device is up to date or not. With manual upgrades, the user will need to look at their profile for the device and ensure that all the software is up to the normal standards. In order to see this, logging into the profile will set up a notification box to say whether or not the device needs to upgrade. This part of the manual upgrade will come automatically. The reason why this happens is to give the user more control over the device to make sure that they know everything going on with the device. All of Microsoft, Windows, and Apple will put these two options into the device for the user to decide. Amazon launched the same type of system for the Amazon Echo. They launched a new and improved system on their operating system to give the user more freedom with their own device. Since they are constantly adding more features to the software for the device, they allow the device to automatically accept updates for the software. It is constantly upgraded for the user to enjoy the latest features that the device has to offer. If the user does not wish for the device to automatically upgrade, then they can change it to manually upgrade within their own personal profile settings.

The fifth thing to realize for the Amazon Echo is that the user needs to speak in order for the voice recognition system to realize that the user is speaking. Since there a voice recognition system has many components, it was designed to have the user speak with their most natural voice possible. There is no need to have a higher voice or a higher tone to speak up. Other devices will have this problem since they will not be upgrading; they will need the user to constantly repeat themselves to ensure high quality, and they confuse homophones. The 'Alexa' software is more advanced, since it only requires the user to speak once for a command and then it has them sit through the answer. The device has a personal setting for natural language installed in the software itself. Basically, the processing systems on this device are more efficient and they can process your voice with more ease and more vigilance than any of the other competitor brands. You say your phrase once and you are ready to hear what Alexa will have to say.

The sixth thing to realize is the cloud processing software. This software allows the device to connect to others while hearing any noise that is happening around the device. This includes any person that is talking in the same room, a door that has opened or closed, the television being on, the dog barking, or someone just being loud. The Amazon Echo can take in all of the sounds and figure out which noise came from the user. It can register which person said 'Alexa' and which person was simply being loud. Even though this happens, it still will only record any noises that come after the word 'Alexa' or 'Amazon' are spoken. Another part of the cloud-based processing is the fact that it can be controlled through a manual operation. This is done through a remote control that comes with the device. The user simply speaks into the remote from a larger distance to allow the device to hear. As the device continues to upgrade and Amazon continues to build more towards the software itself, this cloud-based operation will continue to grow. Keep in mind that the device will still be able to have accessories that can be bought later on. For example, if you buy the Amazon Echo tomorrow and it does not have the remote available for another month, then you can wait. The remote can be bought later on to be paired with the device. Alexa has software that is a large step ahead and it can store your own data without any kind of command. Whenever you tell Alexa to remember a date or a meeting, there is no need to press a 'save' button or tell Alexa to save the information. Instead, it will all be saved together to offer more accuracy and efficiency.

The last thing to consider is the type of hardware that makes up the device itself. This section is for the people who are looking to have the most optimized performance levels available through the processors and hardware components. Since the Amazon Echo is truly a small device that can be held or put anywhere, some people wonder what parts make it into such a powerful device. The main part that powers this device is the DM3725 ARM Cortex-A8 processor. Simply, this is the processor that provides more efficient voice and profile software for the user. The second main part of this device is the 256MB LPDDR1 RAM. This is the component that controls the accuracy for the speed of the device. For the memory, there is also four Gigabytes of storage space for the recordings themselves, the music playlists, and the events that can be marked on the calendar. The 'Alexa' software was created with these main three components to have enough speed and power for a clear voice recognition system. The hardware itself is difficult to compare to the software capabilities. Basically, it is hard to compare the speed and the memory components of the device with the accuracy of the voice recognition software.

All in all, the Amazon Echo is a powerful artificial assistant to be placed in your home. With the capability of searching the web faster to answer any of your questions, playing your favorite music, storing your needed dates, and keeping track of any sports, it is a great little device to have. With the powerful speakers for the music and the efficient voice system, it will not need to be connected to any other part. Requiring the key word to be spoken for a command allows for a maximum amount of privacy for any type of voice recognition software. All of these parts of the device make it the better choice in purchasing to have a faster pace in finding information and therefore more time to relax with your own family.

Simon Monty

Chapter 3
Practical Applications

After reading about all the components that come with the Amazon Echo, you might still be unsure about whether to purchase the device or not, or you might want to read about other ways to use the device. Every person who purchases the Echo knows about the ability to ask hundreds of questions and being able to save scheduled events. Some other people may want to know more about what can be done with the device. While it all seems cool to have a personal assistant, what truly makes this device a multi-purpose piece of technology?

The first basic use is the ability to set either an alarm or a timer. Using the alarm is as simple as giving a command: Say, 'Alexa, wake me up at 8am' to have an alarm set for 8am. Another thing to realize is that the device can be on mute and the alarm will still go off. (You can tell that the device is on mute when the light ring is red). You can set this alarm for any time of the day and you can set it to go off during any day of the week. In order to snooze the alarm, simply say 'Alexa, snooze.' The snooze on the alarm will last about nine minutes before going off again. The alarm can be turned off by saying, 'Alexa, turn off the alarm'. It is important to remember that the alarm will not reset itself automatically. Another important thing to remember is that you are only able to set the alarm 24 hours in advance. Alexa is created to be like a person, and a person will forget an alarm that was supposed to be set later on in the week. The timer on the device will work the same way. Within 24 hours, the user can set the time to go off by giving the same type of voice command for the alarm. For the timer, you are still able to mention the exact number of hours, minutes, and seconds needed. Alexa will accept the command and start the timer. Pausing the timer, though, has to be done through the app. Alexa will still be able to tell the user how much time is left on the timer as long as the user asks. The app itself can be found the same way that any app can be found on a phone.

A much more useful feature of the device is the ability to ask questions and collect any type of necessary information from the user. Since many people can use their smartphones to ask many types of questions, being able to ask Alexa any question is an important component. Whether the question is about cooking, finding a recipe, finding a restaurant, figuring out a math problem, finding a new park to walk around, asking about the year, or other similar things, she can provide answers. In fact, Alexa is even able to tell a joke if the user has the time. She can read information from Wikipedia or even find information from the nearest library. Anything that can be found online can be found through this device, except that the user can get the information faster by being able to say it instead of typing it. In this age of technology, being able to speak a question has become more modern and essential. Since this device has more powerful devices inside to ensure that the user will be heard clearly, it is better to have it for speaking a question when compared side by side to any other kind of smartphone.

Asking any question is as simple as giving a command. The Amazon Echo by itself will be connected to Bluetooth and Wifi to ensure that there is always a strong Internet connection. It follows the same idea of having your own personal assistant at home that can do any form of research and can track any questions that you might have. All the user needs to say is 'Alexa, tell me about the nearest planet,' and the device will find the answer through the closest website on the Internet and then provide you with the top-rated answer based on the other people that browse online. To get a more personalized answer, you can say your zip code or postal code to give a more accurate idea of what the weather will be like around your own location. You can find out the forecast for the entire week or ask about what the weather is like in any other part of the world. Through this command, 'Alexa, what is the weather like today?" the user can find out what parts of the day will be like. Through the use of the Accuweather forecast system online, the user will find out what the weather will be in all parts of the world can be as well as in the user's own locality

Another useful command of this sort is to ask Alexa what the traffic will be like during the day or at that moment. The user can even ask what the traffic is like on a certain road or in a certain city. The best way to approach this is by setting your own starting point and the end point of the destination. From there, the system will act the same way as any other GPS system. It will give you the estimated time it will take to make the trip as well as telling what the traffic will be like in the area. If the user wants to get constant updates from Alexa through the device, they can take the device along with them in the car. If you want Alexa to be the one who is giving you the directions while notifying you of the estimated time of arrival or the lookout for traffic, then simply take the device and plug it into the car.

The third basic command for the Amazon Echo is asking it to manage a shopping list or a to-do list for an event. The first time a user downloads the Amazon Echo app, they can see that adding a shopping list is one of the first things that appears on the screen. From there, you can manage your grocery list while setting your routine for the day. On this grocery list, you can add any items, edit any items, or remove any items. All of it is simple to do. Once the list has been created, Alexa will respond to any type of commands about the list. If the user wants to print out the list however, they will have to use the desktop app instead of using the Amazon Echo itself. The app can also be used to search for different items on sites such as Amazon or Bing. Each item can searched to give the viewer a faster way of seeing the price and the details of the item. The user can see up to 100 items on any lists created through the app. The user will also be able to view them from the Alexa app even when the phone is not connected to the Internet. This means the user can check the list at the store or any place that the phone cannot connect to. The device is then edited through Alexa and the user can control the list through the app. This will make it easier to create the list and to add more items while walking around the home. For example, imagine cleaning the living room when you suddenly realize you are running low on paper towels. You can choose to try remembering it or to write it down on your phone. Pulling out your phone will take more time than simply saying the command, 'Alexa, add paper towels to my shopping list.' This will save time, as you can continue cleaning the room. Imagine you are cooking a meal and you are not able to take a moment to take out your phone and write down a note that you need to buy more salt. Instead of forgetting, simply saying the command, 'Alexa, add salt to my shopping list,' will keep you going in the kitchen without the risk of ruining any food. All lists can be controlled easily. It will save lots of time in the long run as the daily routine can be the main focus with Alexa in the background being able to keep track of any items that the user says that they are running out of in their home.

The fourth basic feature that makes the Amazon Echo stand out even more is the ability to read audiobooks. This is one of the newest features of the device, as Alexa is now able to read any book that has been found through the device. Other devices that can find an audiobook are not fully able to follow through with reading the book itself. Instead, they have a hard time with missing files, causing the device to skip through major parts of any book read. Through the Amazon Echo, that has been changed. Amazon created it to be more user-friendly, so it can find the audiobook faster and it will follow a more consist flow when reading through the material within the audiobook. These new features that have been added do more than playing music through Amazon Music, Amazon Prime, Spotify, etc. Now it can have a stronger Bluetooth connection to add more stability for a reading session that the reader can sit down and listen to. Any media for any audiobook that can be found through a smartphone can be played through the device. The moment you can sit down and have it on your phone, simply pair the devices and then work to relax yourself through the new soothing voice of Alexa.

There is only one limitation that has been found in this device. When listening to an audiobook, there is no control for the playback through voice commands. Playing, pausing, or scrolling through the book, or switch to any other tracks, all need to be done through the phone. There is no command for the device to do these kinds of actions. At the moment, the firmware is being upgraded and enhanced to allow for a better experience for the user. Amazon will then be able to engage this issue and hopefully fix in the near future. The Amazon Echo constantly will need to evolve and be extended. Since it is an assistant to the user, the device will stay up-to-date with any changes that are made. Aside from that one limitation, the Amazon Echo can read the book smoothly so the user can lie down and relax. Instead of having a screen in your face or a book to hold, you can relax on the couch and close your eyes while listening to your new favorite book. Imagine hearing all it being read to you as you paint a picture in your mind of all the events happening. Instead of focusing on reading, you can focus directly on painting the picture. This will put you in the middle of the book with all the action happening. There will be more focus on the details of the book as you close your eyes and listen to the wonders of the new world enveloping you.

The fifth important feature of the Amazon Echo is the ability to order items directly through the device. Earlier, this book described how to create a shopping list or a to-do list through the Alexa app and the device itself. Now imagine ordering an item through the device. You can use Alexa to order an item from Amazon that has previously been purchased. Alexa will then say the name of the product and the price of the item before buying it. If you want the order to be place, say 'Yes' after all the details have been said. All of this will be done through the same shipment and payment methods that are found in your Amazon account. It is all the same policy used with all Amazon orders.

The sixth important feature of the Amazon Echo is t the user can link the device to their own calendar. Since almost any two pieces of technology can be connected to each other, this is another aspect of the same idea. A computer can link to a phone. Now this same connection can happen through the Amazon Echo. It can look through the calendar that has already been set through the phone and then it will be accessed through Alexa. Any other person in the house can link their own calendar to this device. The only limitation is that the calendar must be a google calendar that can be found on most phones today.

The seventh most important feature of this device is that it can completely change your home into a smart home. Whether you have your own house or an apartment, this device can control the entire setup of the place with only a few extra steps. The Echo can integrate the home through home automation hubs. These hubs are from Wink, Insteon, and SmartThings. This is where the user has the choice to buy different light bulbs and controls for anything with a switch or sensor. These can be Philips Hue Lights, Belkin WeMo, Sensi, or Ecobee. There are others that are available online, but those just mentioned are the main ones that can be used to transform the house into a smart home.

These devices together will allow the user to control the lights, temperature, and even video recording within the house. Basically, any electronic device in the house can be controlled through the Amazon Echo. This kind of service is one of the most popular features of the device. Meanwhile, the Echo can still be used to answer any question that comes to mind.

The voice commands for this kind of service are a bit different from the other voice commands, but they still will follow the same type of routine. For example, normally people would think that 'Alexa, turn the lights on' would be the command when walking into a room. Instead the actual command that should be used is 'Alexa, turn the bedroom lights on.' It is the same thought and the same concept. The only change is the detail of the room to ensure the device that the user wants that specific light on. Let us also mention the idea of getting close to bedtime or getting ready to watch a movie. A person will want to begin dimming the lights to make these tasks easier. The command for this would be 'Alexa, turn the (room) lights to fifty percent.' This will make the lights in that room become half as bright. The same idea happens for the temperature control even. Following with the command 'Alexa, turn the heat to (temperature)'. It will then respond. (You will need to have either the Nest or the Ecobee 3 to set up this kind of command for controlling the temperature). The last part of controlling the home is having a video command. It allows for video mode to be turned on to begin recording or watching a show. In order for this to work though, the user will need to get a couple of command triggers from the Belkin WeMo smart switches in the home.

It all might seem to be complicated, but it actually it is pretty simple. An ordinary light bulb cannot be controlled by the Amazon Echo, so it has to be replaced by another light bulb that can be linked to the device. The Echo will then become more than a voice-controlled personal assistant on your kitchen counter; it will be like a remote that can update the entire house. It can help you to control the temperature and turn off the lights while stepping out the door. It can also create a wider picture for the home automation system. There truly are many parts to how far a person can go with this idea. Starting with the light bulbs, they can move all the way down to controlling the sprinkler system in the lawn. All of this can happen through the Amazon Echo.

The eighth most important feature of this device that makes it stand out even more is that it can be used as a kitchen assistant inside of your own home. Alexa can store many recipes or go online to find a recipe to suit the meal. Maybe you are going through the process of cooking chicken but you are not sure about what side dishes to have with it. Maybe you are cooking and forgot a step in the recipe. Maybe you are cooking and something does not seem right about the way that the food is turning out. Instead of panicking or asking a ton of questions at once to waste time, let Alexa be the assistant chef in the kitchen. Letting Alexa read out the recipe will make it easier time to prepare the meal; or you can simply have Alexa remind you of the little steps that you might have forgotten. Either way, having the device read out the recipe will give a more unique time in the kitchen preparing the dishes for the family by having you focus more on the pots rather than staring down at a cookbook.

Not only can Alexa read out the recipe as you go along, she can also help out by converting units of measure. When it comes to switching back and forth between recipes, the recipe will not always match directly with the equipment in the kitchen. For example, most people will have a measuring cup that measures one cup, half a cup, or a quarter of a cup. Some recipes will ask for the amount of water or milk in quarts instead of in cups. Instead of letting the food sit on the heat for too long as you do a Google search to find out how many cups are in a quart, Alexa can make the job much easier. By asking, 'Alexa, how many cups are in a quart?' she will tell you and you can keep on cooking. The next great part about having Alexa right in the kitchen is the ability to use her as a timer. This was mentioned earlier, but it is being said again since timers are used a lot in cooking. With all of help in the kitchen with this little device, there are many other ways to explore cooking. Just being able to throw food in a pan and leave it after telling her to set a timer will make life in the kitchen much better and faster.

Another aspect to having Alexa in the kitchen is having a natural calorie counter. This can help anyone who wants to keep track of their health or diet. It is simply a way to keep track of calories in all the food that is being cooked or eaten. By saying 'Alexa, document one turkey sandwich and one apple,' she will tell you how many calories are in that whole meal. From there, she will keep that information saved inside of the system to help you compare and keep track of the diet. This will make it easier to document since you will not need to research every part of the serving size or the amount of calories in one part of the food. Alexa will research all of the information together to provide a more accurate representation of how many calories are on your plate.

The ninth important feature of the Amazon Echo that is important to know about is ordering things through the device. Since it is all connected to the Internet, anything that can be ordered online can be ordered through this system. These kinds of orders can include getting transportation, ordering pizza, or even ordering flowers. Whatever can be ordered online can be found through the device.

Two main places that can be reached through this device are Uber and Domino's, which are among the more popular places for phone apps. A person can order different types of pizza and a different form of transportation. The other popular item to be ordered is flowers. This service mainly comes from 1-800-flowers. It is another case where the service is used the most through the Internet, and therefore people are going to refer to it for the needed flowers or decorations. Any kind of service that can be found online can be used by this device. Even though there are these main types of services that have places paired with them, they will still continue through to create a larger picture for people to follow.

The final important feature is that Alexa can play games and tell jokes. When it comes to telling a joke, be prepared to hear a rather dry or corny joke. If that's are your thing, you will enjoy hearing these jokes over and over again. Otherwise, this system of jokes is to help the user get a quick laugh in the middle of their day. If you are in a rush and need a quick break or something to remember on the way up to the office, then a joke can be just enough to help you get through the tasks that fill up the rest of your day. One small joke, whether good or bad, can go a very long way to show a person a reason to continue moving forward.

The other part of this feature in the Alexa system is the ability to play games that can be found on the Internet. These are not the types of games that are heavy with visual effects or single-person games. Instead, they are games that require two players. For example, bingo is one of the games that can be played through the device. It is a two-player game that is simple to follow and the device can be that second person. Other games on this device include Simon Says and Jeopardy. These are just three of the games that the Amazon Echo can play. They only need to be the kind of games that involve two players that have more of a creative side, rather than having the game guide them.

The device can also offer fun quizzes for people to try. These quizzes include 'capital,' 'star wars,' and other quizzes that can be found online. Whether it is about what character from a movie they are most like, what food fits their kind of personality, or what mental age they actually are, all kinds of quizzes are out there to ensure that you never get bored. If you ever have a bored moment with this device and you are trying to think about what kind of game to play, one of the really fun ones to play is called Word Master. This is a game where a letter is said and people have to say words that begin with that letter. It is a way to test your own level of knowledge to see how many words you actually know while seeing what parts of the dictionary you are most familiar with. Either way, these two things provide hours upon hours of fun through laughs and jokes for the whole group of friends and family to be a part of.

These are only the ten main important features of the Amazon Echo. There is still more to be discovered. Since the device continues being updated constantly, Amazon tries to ensure the most satisfaction for all of its users. As a result, this is a great product to have within your own home.

Imagine being able to walk in and have the device control the lights by a simple command. Imagine being able to have the device read your favorite book as you relax on the couch and stare at the ceiling. Imagine being able to run around cooking and cleaning while building your own grocery list. All of this is possible with this new device. Even though it seems to be rather small, it is still a powerful; device that can import many different formats. Being able to connect to the phone while also being able to connect to the computer is possible because this device runs on Bluetooth. Since there is a lot to see and many things to do, this device helps users to stay on track with its new ability of following a schedule. Saying how the weather is, how the traffic is, and how the daily schedule are among the features that make people want to buy the Amazon Echo.

Chapter 4
Using the Amazon Echo

With all the different benefits to explore when using the Amazon Echo, it may seem rather strange when you think about how to set it up. While it might seem simple to turn it on and connect it immediately to the Internet or to Bluetooth, it still might be overwhelming to think about all the different commands and steps that a person needs to go through to completely set up and personalize the device. Normally, the instructions will be enough to go back to in order to have a general idea of how to set something up. Before purchasing though, it might be difficult to find the right instruction manual or anything that can describe how to work each feature on the Amazon Echo. If you read this chapter, there is no need to look any further in researching online. The following parts of the chapter will list a majority of the different features in how to use them.

The features listed below are the more popular features that were mentioned in the benefits of this device. Since these are widely known, they are written out with more detail about how to use those features and how to work with Alexa to set up those parts of the device. Amazon has worked hard to develop this kind of software to be more friendly to the user. This will allow for an easier and faster time setting it up. Whether you want to set up your home to be a smart home, set up a timer in the kitchen, set an alarm to wake up, or even see how to ask a question, all of it will be listed below to ensure that you know the benefits of owning this device.

Overall Setup

This is the moment when you have ordered the Amazon Echo online and the box is sitting right in front of you on the table. Once you open up the box, you see all the parts that come with it: The power cord, the device itself, two batteries, and the remote control. The first thing to do is plug the device into the wall using the power cord. Once this is done, insert the two batteries into the remote control. The moment the batteries go into the remote control, it will turn on and be paired with the Amazon Echo.

If you are not able to get the remote to pair up with the Amazon Echo the first time, there is no need to get stressed out. The thing to do at this point is to open up the Alexa app and follow the instructions after downloading it. This will allow for the app to find the device after it has connected to the house and item number of the device. The app can be downloaded on any Android phone or tablet, any iOS phone or tablet, any Amazon Fire Phone or Fire Tablet. The app can be found through the Google Play store app that is normally found on the front screen of the devices listed. If it cannot be found there, then look online, using the same search term to find the app. Once the app has been successfully downloaded, go to settings and select the Amazon Echo. Then choose the option to pair remote. Now the devices should be fully paired together.

With the remote control and the device paired together, the next step is to connect the device to the home Wi-Fi network. This is done after plugging the device into the wall. It will sense if there is any kind of Wi-Fi that can be connected to and it will try to connect. If the Internet in the home is password protected, the user will need to go through the app to set up the Internet connection. It is important to remember that the Amazon Echo will not support any enterprise or ad-hoc networks involved with the Wi-Fi. It must be the kind of Internet that can be used at home. It cannot be the Internet that comes through a company Intranet.

If this way of connecting to the Internet does not work for the user, then they will need to connect to the Internet through the Alexa app. After opening the app, click on settings, and find the 'Set up new Echo' option. Click on that button and hold down the action button on the Amazon Echo for about five seconds. The circular light around the top of the device will turn orange as the phone connects to the Echo. A list of the Wi-Fi networks around that area will appear on the front screen. Find your own Internet network, and click on 'connect.' If you are not able to find your own Internet, then click on 'add a network' or 'rescan' to find the Internet network manually. Once the network is connected, there will be a confirmation message that appears on the front screen of the app. You can then return to the home screen.

The software of Alexa works on the concept of cloud computing. This means that the user can access the

Amazon Cloud through their own Internet. Simply open up the app and follow the instructions to find the picture of the video that you want to see. Following the instructions, there will be an instructional video that shows the user how to use the basic element of the voice commands on the device. It will also show the user what else needs to be done in order to get the Alexa device up and running.

In order to double-check that the device is connected to both the Wi-Fi network in your home and the Cloud, first check the power LED light on the top of the power cord. Either way, there will be a light there. If the light is a solid white color, then the Amazon Echo is fully connected to the Wi-Fi network and the cloud. If the light is a solid orange color, then the device is not connected to the Wi-Fi. If the light is an orange color and blinking, then the device is connected to the Wi-Fi, but it is not connected to the cloud. In all of these cases, there will always be a light on the top of the power cord. Depending on what the color is and what the color is doing will tell you what is happening with the device.

To fix any problems that involve the device not connecting to the Internet, go back to the Alexa app. To start, find the original Wi-Fi network on the app and try to disconnect it and reconnect it. If this does not work, then reset your router. If neither of these is working, double-check to see if you chose the correct account for the Internet or if you typed in the password correctly. If none of these options work, then try to move the Amazon Echo closer to the Internet router. Sometimes it needs to be nearer to the Wi-Fi signal in order to work. Once it is closer to the router, repeat the steps above to try reconnecting to the Internet once again. If problems still occur, try unplugging the Amazon Echo for about fifteen seconds; then plug the power cord back in and see if that fixed the problem. Another thing to double-check is whether your Amazon Echo is registered on your Amazon account. In order to see if it is registered or not, log into your Amazon account and go to 'Manage your content and devices.' From there, select 'your devices' and try to find the name of your device. If it is registered and still not connecting, then deregister the device online and log out of the account. Try the process of setting up the Amazon Echo again to register the device a second time. If none of the above options works, you will need to contact Amazon Help and Support to see what the problem actually is. Since every device is different and every situation is different, these experts will guide you to find out what the problem is and how to find the solution.

After all of these steps have been completed and the Amazon Echo has fully completed the setup, it is time to check the setup. The first thing to do is to give it a name in order to have it identified through the software. If you have multiple Amazon Echo devices in your home, giving each one a name will help to keep things clear. Each device will need to be called by and identified through its name. In order to change the name on the Amazon Echo device, open the Alexa app and go over to settings. At first, the device that you just purchased will be called 'Your Amazon Echo'. Click on the name field to select the entire default name, delete it, and then type in any name that you want to give the Echo. Once a name has been put into the app, click 'Save changes' and go back to the home screen. Now the device will have a name of its own. It is also important to remember that this is still only the name of the overall device. This name will not act as a wake word for the device to respond to. Instead, the name will help to keep the devices controlled through the Alexa app.

For the wake word, many people will choose to stay with the name Alexa to continue with the overall theme of the device itself. The only names that can be used as the wake word are Amazon, Alexa, or Simon. These names will be used whenever the user wishes to ask a question or give a command to the device. The device name will be used to connect the device to Bluetooth, the app, or any other network outside of the Amazon Echo.

The last part of setting up the Amazon Echo is about making the device a bit more personal. This is where you can define any specific personal specifications for the device. By personalizing the device, Alexa can see what your own preferences are and it will make the software a bit more capable of becoming your personal assistant. The first step in this part of the process is to input the location of the Amazon Echo. In order to do this, log on to the Alexa app once again and click on 'Settings' on the left side of the navigation panel. From there, you will see a 'Device location' tab on the screen. Click on it and then enter your zip code. If you see that there is already a zip code in the field, this is because the software will automatically attempt to find the location to see any Wi-Fi points in the area.

Check to see if the zip code is correct and then leave the number there. This part of the setup is not available to any people that are living outside the United States of America. If you live in one of these other countries, you will need to enter your own postal code manually into the Alexa app. No matter which part you are on, be sure that the code is correct. The device will then be able to tune into any other radio stations that are in the same area, get any news about the local weather close to your home, and will be able to do localized tasks for you.

The second step in personalizing the Amazon Echo is to select the system of units that you are most comfortable using or hearing about. This is about how the device will choose to say, for example, miles or kilometers. You can choose between the metric system or the standard system of units for measuring the different distances or between Fahrenheit and Celsius for temperatures. To change any of these settings, go into the Alexa app and click on 'Settings' on the left side of the navigation panel. From there, pick the style of units that you want to work with whenever using this device. Alexa will then always respond to your questions through the system you put into the app.

The third step in personalizing the Amazon Echo is the Amazon account settings. This is even easier than the other steps if you have already successfully registered the device on your computer. Once it is completely registered, you can manage any account settings through the Alexa app. The most important feature in this step in the voice purchasing feature found online with the help of Alexa. Simply give her the command for the item you want to purchase and she will do the rest. Everything in this part of the voice purchasing option is explained later on in this same chapter under the 'Set up placing orders to stores or Amazon.' Users can manage their one-click preferences through this part of the app. It makes the process of buying a product much faster and much easier.

With the ability to handle multiple user accounts while also being able to provide users with this many features makes the Alexa software stand out above its competitors. This device can give a new perspective on how to live through your daily life through all of these simple steps. There are many more features that can be explored through the Alexa app when using the device, but these are the main setup options to get started.

Light Ring

There is a light ring on the top of the Echo that shows the user what is happening with the device, as a phone that has a red or white light in the top corner. The light might blink slowly for an email and faster for a message, but overall it shows the user that something is happening with the device. The user only needs to see what color is showing to make sure that they do the right thing. All of the colors are very different from one another, and therefore there will be very little confusion about what the meaning is. Below is the list of the different colors that can appear and what each of these colors means.

-<u>Solid blue with a spinning cyan light:</u> This colors means that the device is beginning to start up. It has just turned on and it is beginning to load. While it is spinning, it is not yet active, but it is preparing to be active.

-<u>All lights off:</u> This means that the device is finally active and is now waiting for the next request or command.

-<u>Solid blue with a cyan color pointing in a direction:</u> This means that the cyan color will continue pointing and moving towards the person speaking the request or the command. It is attempting to process the person's request, and it points to show the focus on the person who is saying the command. Otherwise, no one would know which person the device was listening too.

-Orange light that is spinning clockwise: This means that the device is trying to connect to the Wi-Fi network.

-Solid red light: This means that the device has all the microphones turned off and that they were turned off through the top button on the top of the device with a microphone symbol. The way to turn the microphone back on is too press that button again.

-White light: This means that the user is adjusting the volume level on the device.

-Continuously oscillating violet light: This means that there was an error in attempting to connect to the Wi-Fi.

Remote Control
The Amazon Echo comes with a remote to help with navigation and use of the device. It allows the user to perform all of the different functions for the device without the use of their voice. This can be helpful when they need to

be quiet or if they are too far away to be heard by the device. The remote design is very close to the looks of the Amazon Fire TV remote control. It has a rubber grip to add comfort and support and a microphone is built in at the top of the device to allow the user to talk directly to Alexa without having to shout from a longer distance.

To set this up, all the user has to do is press the power button on the top of the remote while making sure that the Amazon Echo is turned on. With both of them powered on, they will automatically connect as long as they are within range of each other. Through this function, the user can control the volume of the music, the playlist of songs, and other settings that come with the device. Another interesting part of this remote is that it has a magnetic holder attached. This allows for the user to place the remote on any magnetic surface instead of having to carry the remote around throughout the house.

Set Up GPS for Traffic or Directions

In order to set up the travel information for a trip, the first thing the user must do is open up the Alexa app and look to the left of the navigation panel. From there, click on 'Settings,' and then the 'Traffic' button. The next step is to the addresses in the 'from' and 'to' dialog boxes. Then click 'Save changes' to save the two locations for your next trip. This will store the information needed for the device to say what traffic there is and it can help the driver navigate. It acts the same way as any other smartphone or GPS device. If a person wants to add in another stop on their route, then they click the button 'New stop' and add the details of the next stop. To ask or give Alexa the command for the kind of traffic updates on the route you have chosen, say 'Alexa, how is the traffic?', 'Alexa, what is the traffic like now?', or 'Alexa, what is my commute.' It is all very simple to follow and set up, as Alexa will guide the user.

Set Up a Shopping List or To-Do List
In order to set up a shopping list, you should first realize that many parts can be done either through the Alexa app or through a command, but not all of them can be done both ways. The first part of the dash under each important step will be for the use of the Alexa app. The second part of the dash will be for the command that can be used. To begin, here's how to add an item to the shopping list or the to-do list:

-Through the Alexa app, open up the shopping or to do-list from the left part of the navigation panel. Then type in the name of the item you want to add to the list and then click the '+' sign.

-Once the list is created, a command can be used to add items to the list. The command can take several forms: 'Alexa, add (item name) to my shopping list,' 'Alexa, I need to buy (item name),' 'Alexa, I need to (chore, hobby, meeting, action, etc),' or 'Alexa, put (chore, hobby, meeting, action, etc) on my to-do list'.

Here's how to review your shopping list or to-do list.

-Through the Alexa app, open the shopping list or the to-do list from the left part of the navigation panel. From there, you can read all the items on the shopping list.

-To use a command, you only need to say 'Alexa, what is on my shopping list?' or 'Alexa, what is on my to-do list?' to access any part of the list. It will be read to them through Alexa.

Here's how to edit items on the list.

-Through the Alexa app, open up the shopping list or the to-do list, click on the item that you want to change, and type in the changes.

-There is no command for this action.

Here's how to remov an item from either of the lists.

-Through the Alexa app, open up the shopping list or the to-do list, click on the item that you want to remove, click the down arrow beside the item name, and click 'Delete item,' Another way to delete an item is too tap the checkbox by the item and click on 'Delete' in the app itself.

-There is no command for this action.

Here's how to view any tasks that have already been completed.

-Through the Alexa app, and choose the list that you want to look at through the left part of the navigation panel. Click on the 'View completed' to see the list of completed items. To remove all of the items from the list, click 'Delete all.'

-There is no command for this action.

Printing a shopping list or a to-do list from the app can be only through the Alexa app that can be found with either a computer or a phone. You need access to a web browser and a printer to do this.

-Through the Alexa app, find the list that you want to print from the navigation panel. Click on the button 'Print' on the top right corner to print it out. Be sure to check if there is a connection to the printer.

-There is no command for this action.

The last important action is searching Bing or Amazon for an item on the shopping list.

-Through the Alexa app, open up the shopping list, select the item that you are searching for, and click on 'Search Amazon for' or 'Search Bing for'. From there, the app will show you the prices of the item that you want to buy.

-There is no command for this action.

All of these actions with the shopping list are easier on the app than on the device itself. Even though the lists are mainly controlled through the Alexa app, the device still stores the list and reads it to you. Once the list has been created through the app, the device can be used to add items. Otherwise, the user can go through their daily routine while speaking the next items that they want to add to the list.

Set Up an Audiobook

In order to set up the audiobook feature of this device, you first have to have an audiobook to read. Whether you have one in mind or you are simply open to any option, you need to fine the book online. Search for the specific book or your favorite genre. The Amazon website offers a fast way to find and purchase the book. You can also use the Amazon website to find long lists of thousands upon thousands of books to choose from. If you have your own Audible account, it can be used for this function, too. The eBooks found in the Audible library can be connected to the device the same way.

Once the book has been bought and stored on the phone, it is simply a matter or pairing the two devices to allow them to connect. The first step is to set up the Amazon Echo as a Bluetooth speaker. This can be done by saying the command 'Alexa, pair Bluetooth.' Alexa will give you the instructions on pairing the phone to the device. Doing this will require going to the Bluetooth settings in the tablet or the smartphone and then pairing it to the device named 'Alexa.' After this has been done and the two devices are completely paired, you can enjoy having the audiobook read to you through the Amazon Echo. From here, all you need to do is say the command, 'Alexa, play the audiobook (book name),' and it will begin reading. Audible itself has begun to download their own audiobooks to the library free for their own Amazon Echo users to enjoy.

To begin using this function through Audible, you need to download the app. To have the Amazon Echo begin reading the eBook found in the Audible library, simply say, 'Alexa, read (book name)'. You can pause the book or stop the book through the phone. The device will save the place that the book was stopped for later on. You can then resume reading the book from the place you stopped by saying, 'Alexa, read my book.' Later on, the playback functions should be control through the device when it is updated. For now, you have to rewind the book or scroll through it by using your smartphone.

It is truly that simple. Once the book is bought and the Bluetooth is set up between the device and the phone, the book can be read through the device with ease. Other devices can do the same, but only through the full control of the display. Amazon was able to take it one step farther to allow for more ease for the user. It brings a new way to listen to books through a voice-controlled device. You can now lie down and listen to Alexa read you a story from your new favorite book.

Set Up Placing Orders to Stores or Amazon

In order to set up placing orders through the device, you must first have an account with Amazon. Not only that, but only items that were purchased from Amazon through your computer or phone can be purchased again through the Amazon Echo. To place the order, say, 'Alexa, reorder (item name).' Once this is said, Alexa will say the name of the product and its price. In the Alexa app, more information about the product can be seen. To confirm the order, say 'Yes' and the order will then be placed. It is the same process that is used when ordering an item from the computer. It uses the same shipping and payment information that were used on the product previously. They are covered by the same free shipping policy that Amazon is widely known for.

Another way to make changes to the method of ordering products is by going through the Alexa app. For example, if you want to turn off the voice purchasing option, you can set up a voice PIN code. This code will have to be used any time you want to make a purchase. This will add more security by preventing any other person from using your your account and it will also stop any accidental purchases. To set up the code, open up the Alexa app, go to the left of the navigation panel, and select 'Settings.' Choose the option to activate or deactivate the voice purchasing option. Here you will also be given the choice of setting up a PIN code to add security to the device.

If the item that you're trying to purchase is no longer in stock or no longer sold through Amazon, the Amazon Echo will provide you with other choices through Amazon's Choice. In this system, the device will find the products that are sold by other and will choose the item to be mentioned first based on both the highest rating and the price of shipping. It will still be the same item, it will only be done through a different seller besides Amazon. This will not happen often, but it is possible. Another possibility is that the item is not available at all through Amazon or any other buyers.

To begin ordering your items through the device, it is important to know all the different commands that can be used. As with creating the shopping list or the to-do list through both the Alexa app and the device itself, purchasing a previous item through the device can be done only through the commands on the device.

To start, say, 'Alexa, reorder (item name).' Depending on with the status of the item, the Amazon Echo will give a different response. The first possible response is when the Echo finds that there is a previous order for that item.

-If you do not have a confirmation code, then Alexa will say 'The order total is $(amount of order). Should I order it?' You replay 'Yes' or 'No.'

-If you created a PIN, simply say it after Alexa says the same thing mentioned above.

The next problem that might come up is that no previous order for the item was found. When this happens, the device will find an alternative way to purchase the device through another buyer.

-Alexa will then say 'I did not find that item in your previous order history, but Amazon's Choice for (item name) is (price of item). The order total is (total price of order). Shall I order it?' Then say either 'Yes' or 'No' or your PIN code to confirm the purchase of the item.

If the item cannot be found on Amazon or through another seller, the Echo will notify you and put the item in the shopping list.

-Alexa will say, 'I did not find that in your past orders so I have added (item name) to your shopping list.'

The next problem that might come up is that the item that you are trying to order again is not available any more as an Amazon Eligible item.

-Alexa will say 'I found (item name) but I can only order Prime Eligible products. See your Alexa app for options.' Open the app the see where to find the item again or to purchase the item through your phone.

The next problem that might come up is that the item is no longer in stock.

-Alexa will say 'I found (item name) but it is temporarily out of stock. See your Alexa App for options.' Open the app and save the item to your wish list.

The next possible problem is that the item is eligible for Prime shipping, but it can only be bought as an add-on item.

-Alexa will say 'I found (item name) but I can only reorder products for Prime members.' Go onto the Alexa app and see what other products will need to be bought to ensure that the product can actually be purchased.

Another problem that might arise is that you are trying to place an order using an account that does not have an Amazon Prime membership.

-Alexa will say 'I found (item name) but I can only reorder products for Prime members.' Go onto the Alexa app and find out how to become a prime member. Otherwise, you can order the product only through the computer or phone.

The next problem that might happen is that you will place an order that has an issue with the billing address on the account.

-Alexa will say 'Sorry, but there is a problem with the billing address on your account.' You will need to go to Amazon.com to see what errors are occurring and then complete the order.

When placing any order through the device, it might be possible for the device to see that you are trying to place an order, but the order might not go through. Alexa will still be able to tell the user that there is a problem and they can say what needs to be done to fix the issue. Most of the time, if there is a problem ordering the item through the Echo, the user will need to go onto a computer or the Alexa app to order the product through there. Otherwise, it is as simple as saying another command to have another product ordered through the Amazon account.

Set Up a Calendar
In order to link the device to a calendar from the phone, it has to be a Google calendar. To begin linking the Google calendar to the Amazon Echo, open the Alexa app. Go to

the left of the navigation panel and find 'Settings.' From there, go to 'Calendar' and click on 'Link Google calendar account'. Use the login details for your Google account and follow the instructions. When you've finished all the instructions, the Amazon Echo will have access to the calendar.

To manage your calendars, open the Alexa app. Go to 'Settings.' Click on 'Calendar' and check the boxes next to the calendars that you want Alexa to read out to you. Once you have gone completely through this process, there are more commands you can use to manage daily or weekly schedule. For example, asking 'Alexa, when is my next event?' will give the device the command to go to the calendar and say when the next event is. Then you can see how much time you have before that event comes up. Other commands include 'Alexa, what is on my calendar?', 'Alexa, what is on (name of another person)'s calendar?', or 'Alexa, what is on my calendar (day) at (time)?' All of these commands make it easier to navigate a heavy schedule. Instead of focusing hard on what might need to be done, this makes it easier to have reminders and schedules spoken to them after a simple command.

Managing Multiple Accounts

Since there can be multiple Amazon Echo devices in a single home, there is the possibility of also having multiple accounts in the same household, as well. Since the Echo is mainly a home device, there will be one main account that Alexa will choose to refer too without any details being added into the command. If you do not want to be the only person that can use the device, then other accounts can be added to the same device. With Alexa, the main user can make a Household Profile section for the device to use that allows for any other family members to create their own to-do lists or music libraries. Anything that the user can do, the other family members would be able to do as well. They will even be able to make other forms of joint purchases or they can use the family account that is found through the Amazon device.

To begin setting this up or to begin using this feature, the user must open up the Alexa app and go to 'Settings' on the left side of the navigation panel. From there, they need to select the 'Household profiles' option. Here, the user can add a person to the Amazon household. The really important thing to remember is that, when adding a person to this part of the device, they will need to be present at that moment. The personal details and credentials of the person will be added by following the instructions in the app and then they can have their own spot on the device. After adding in the person, remember to save the person into the account. After that, Alexa can manage two user accounts at the same time.

Since only one account can be running at one time, you need to which one is being used. Simply say, 'Alexa, which account is this?' If you need to change accounts, give the command 'Alexa, switch accounts.' There are still some common elements that do exist between multiple accounts. The shopping list or the to-do list can be viewed by any person in the household. There is no need to switch accounts in order to see your own list that was created through your own account.

If you need to remove a person from the household account, go back to the Alexa app and click on 'Settings' and the 'Household Profiles' tab, then click on 'Manage Your Amazon Household.' Look through the list to find the person that you wish to remove. Select that person and click the 'Remove' option on the screen. To remove yourself from the Amazon Household, click the 'leave' button and then click on 'Remove from household' before returning to the main screen to finish the removal process completely.

One important thing to remember when considering the Amazon Household option for the device is that any person is in the household list will have access to the billing and credit card information. Since this information is registered to your own account and then in the Household Profile, it will be available to any other person who is on the profile. This is where the idea of having a confirmation PIN when purchasing anything through this device will come in handy. If you are worried about someone else spending money on your own credit card or on your own account, then all you would have to do is set up the PIN and keep the number to yourself. Without the correct PIN, Alexa will not purchase any items through the device. This will prevent any items from being purchased without your permission

Final Steps

With all of this set up, all you will need to do is scroll through the app to ensure that it fits your personality. Make sure that the zip code is correct and that the device can register your voice. After following all the steps and setting up everything, the device can be completely used.

Chapter 5
Tips and Tricks

While we've explained many features of the Amazon Echo, there are still many others. Whenever someone gets a piece of technology, the first thing they realize is that there are many features to learn about. They can see that there is an instruction manual that shows the basics on how to use the device. Then, as they use it, they realize that there are certain tips and tricks on how to use it a little bit better. For example, when you're talking on a telephone you can hold the phone up to your ear, but some people figured out that you can use a headset instead. The same idea continues on with the Amazon Echo because users are trying to figure out a way to not only how to update the technology but how to make it a little bit more interactive.

The first place where it needs to be a little more interactive is with the wake word. The Amazon Echo requires a wake word before obeying a command. It does not matter what the command is, a wake word has to be said. Some users want to talk more freely without worrying about whether or not Alexa will wake up and not. The idea is that you want the device to stop listening constantly for that certain wake word. Simply the best way to do this is to just press the mute button that is on the top of the device. Once you press it, then the light ring with be red on the top of it to let you know that the device has been muted. Alexa will then go quiet and will stay muted so that you can talk freely about the device or anything about Amazon without worrying about the device always waking back up. For example, if the device is sitting on the counter and you're talking about its benefits, Alexa might hear the wake word a dozen times over. But if you are describing the device you do not want to be interrupted in that way, since you want to try to get the point across that it is a good device to buy instead of a device that keeps needing to be heard. The best way to stop this is to simply to press the mute button that is on the top of the device.

The second trick that people found was about the Amazon Echo is the updated software. While many people reading that the Amazon Echo is naturally updated software, sometimes the software will not match up on your own device and the upgrade will not necessarily happen. Just as with any other kind of digital piece of hardware, this CPU and the hard drive may not match up and therefore it should be manually updated. It is not something that will happen every single time, but it could happen every once in a while and it is worth looking into to figure out how to manually update the device. This also helps make sure that your device is up-to-date and is also properly working with no errors. There are two ways to manually update the device software on the Amazon Echo. The first way is to use the mute button, which is found on the top of the device. Press the mute button wait 30 minutes. When you press the mute button again, the device will be updated. The second way is to open up the Alexa app and to go into 'Settings.' Then find updates and look at the the list to see when the echo has been updated recently. Mainly what will happen is that there will be a pop-up to show if any updates need to be applied.

The third tip and trick that can be used with the Amazon Echo is learning how to use the web in order to access the device. When you are first setting up the Echo, you have to do it through the Alexa app. This is for setting up the

Internet connection and the Bluetooth connection to the cloud to ensure that all the accounts are fully registered. But you can also any web browser on a desktop computer to find more settings for this device. Not only that, but through the website it is sometimes actually easier to manage your shopping list, with some other forms of the to-do lists. It is because every person has a preference about how they write the lists and about the word screen size and how to gather information. One person might enjoy looking at Amazon on their phone while another person might enjoy looking at Amazon through a desktop computer. Either way the main thing to get here is that you can use a website in order to access all the features of the Amazon Echo.

The next tip and trick is about the family accounts other persons who are on the account. Earlier you read about how to add a person to the account and how to remove a person from the account. This is more about controlling the family accounts that are connected to Amazon Prime. Since this is a shared member household, there is a sense of security that needs to be instilled through the device. Normally people just choose to have a PIN code in order to have a safer way to go with shopping, but if one person truly wanted to go on to the other person's Amazon account, it is still possible. Simply by telling Alexa that you need to switch profiles is the first way to switch profiles. Let us say that a person has five profiles on the list with this device. This means that when you say the command 'Alexa, switch profiles,' Alexa actually goes to the next profile on the list. If you want to choose a more specific profile for Alexa go to, then you say the name of that profile.

The next one is about using this device to turn your home into a smart home. As mentioned earlier, the best thing to do for this is to replace many items in your home with new kinds of technology that can have sensors that will connect to the device. The most common way to do this is with all the light bulbs in the house. After changing all the light bulbs, there will be sensors in each room and the device can be programmed directly to those sensors. For example, as mentioned earlier, the light bulbs inside a bedroom can be programmed so the device recognizes them as the bedroom lights. You can then tell Alexa to turn off the bedroom lights and they will be turned off. Another part of this that is very helpful, especially for the family that has young children, is the ability to turn off the television. This feature contains a little Easter egg: If the person says 'Alexa, Simon Says' and then they say the command that they want, it actually will go through. This can be used to get children in bed on time because then they can go and have the television turned off. Another fun way to play around with the Amazon Echo is when it comes to saying a bad word. Since we do not want children to know these bad words, the Amazon Echo will actually put a beep in the place of any swear word. While this is just for fun, it shows that the language is not going to be used in front of children.

The next tip and trick is about simple calculations and math. The user can ask any kind of math question and Alexa can answer it. Just as Alexa can be used as a calorie counter for anyone on a diet, the same kind of math skills can be used as a calculator. For example, when sitting around the dinner table having a discussion about the distance marks between a mile and a kilometer, Alexa would not only be able to do the conversation, but she would be able to keep track of the numbers. Another example is when keeping a score sheet: Alexa can keep track of each person's score as they play a game. Otherwise, the ability to act the same way as a calculator can be used for homework or accounting help by any person in the family.

This last part comes from an online list of a multitude of Easter egg items for the Amazon Echo. It is a list that is completely for fun that anyone who owns this device should try out. The list is good for those times when you are entertaining guests or simply needing a laugh. All of these things produce a clever joke that any person can appreciate. For all of these, say 'Alexa,' then the phrase that follows.

-I am your father

-define rock, paper, scissors, lizard, Spock

-who lives in a pineapple under the sea?

-Romeo, Romeo, wherefore art thou, Romeo?

-what is the loneliest number?

-how much is that doggie in the window?

-how many roads must a man walk down?

Simon Monty

-all your base are belong to us

-beam me up

-how much wood would a woodchuck chuck if a woodchuck could chuck wood?

-define supercalifragilisticexpialidocious

-who's your daddy?

-Earl Grey. Hot

-what is the meaning of life?

-what does the Earth weigh?

-when is the end of the world?

-is there a Santa?

-make me a sandwich

-what is the best tablet?

-what is your favorite color?

-what is your quest?

-who won best actor Oscar in 1973?

-what is the airspeed velocity of an unladen swallow?

-where do babies come from?

-do you have a boyfriend?

-which comes first: the chicken or the egg?

-may the force be with you

-do aliens exist?

-how many licks does it take to get to the center of a tootsie pop?

-what are you going to do today?

-where do you live?

-do you want to build a snowman?

-do you really want to hurt me?

-what is love?

-who is the real slim shady?

-who let the dogs out?

-open the pod bay doors

Simon Monty

-surely you can't be serious

-to be or not to be?

-who is the fairest of them all?

-who loves ya, baby?

-who you gonna call?

-who is the walrus

-do you have any brothers or sisters?

-do you know the muffin man?

-how much do you weigh?

-how tall are you?

-where are you from?

-do you want to fight?

-do you want to play a game?

-I think you're funny.

-where in the world is Carmen Sandiego

-where's Waldo?

-do you know the way to San Jose?

-where have all the flowers gone?

-what's in a name?

-what does the fox say? (ask a few times, you will get multiple answers)

-Alexa, when am I going to die?

-I want the truth!

-make me breakfast.

-why did the chicken cross the road?

-where are my keys? (ask the question twice)

-can you give me some money? (ask the question twice)

-knock knock.

-what are you wearing?

-party time!

-party on, Wayne.

-is the cake a lie?

-are you sky net?

-your mother was a hamster.

-set phasers to kill.

-roll a die.

-random number between x and y.

-random fact.

-tell me a joke.

-heads or tails?

-mac or pc?

-show me the money.

- what is the sound of one hand clapping?

-give me a hug.

-are you lying?

-my name is Inigo Montoya

-how many angels can dance on the head of a pin? (There are three answers to this one.)

-see you later, alligator.

Not all of these phrases work for every person. They are all from movies, books, or very popular things on the Internet. Even though not everyone will know what all of these are, they still are a lot of fun to play around with and they will get a very funny answer each and every time. Other than that, all of these tips and tricks are very helpful to keep in mind when using your device. If you want to have the best time and get the most for the money you spent on this product, then you want to use the Amazon Echo to its fullest potential. This includes being able to joke around with her, being able to set up any calendar, being able to look up any item, or even being able to just monitor the home. Having this device inside your house is going to make life different, just like getting a new phone or a new computer. Your daily life will stay mainly the same, but it will change a little bit because you just bought something new and you want to have a new sense of excitement and a little bit of a change in your life to bring a spark of joy.

Simon Monty

Chapter 6
Frequently Asked Questions

The first question that is often asked about this device is whether the voice purchasing option can be turned completely off. The answer to this is 'Yes.' All you need to do is open up the Alexa app on your smartphone or tablet, go to the 'Settings' option on the left side of the navigation panel, and click on 'Voice purchasing'. Turn it off, but remember to save the changes before exiting the webpage of the app. Other than that, the only other way is to enter a verification code for voice purchasing that Alexa will then ask for whenever a person tries to buy something through the Amazon Echo.

The second question is, 'How do I purchase music on Echo?' Purchasing music on your Amazon Echo is actually very simple and takes no time at all. It goes even faster if you have one-click payment enabled. All would need to do is place an order and Alexa will use the default payment settings from your account to purchase it. She can also read you back the product details and then you can also use a confirmation code while purchasing to add even more security to your transactions.

The third question that is asked is, 'What happens to deleted voice recordings?' The device can provide you with two options for deleting voice recordings. You can delete them from your device, but they will be kept as a backup in the Amazon Cloud. The second option if you want them completely gone is too delete them both from the Cloud and the device itself. The Amazon Echo will simply remove the date and then the Home Screen Cards associated with each of the recordings.

The fourth question often asked is, 'Can all of my voice recordings be deleted at once?' The answer to this question is a simple 'Yes.' Just to go Amazon where the account and the device are registered, click on the Amazon Echo used for your home and delete them all yourself. The other option is to contact customer support and have them delete everything. The request to delete all of it might take a while to be completely processed, so remember that all of the conversations that are stored in Alexa are completely safe and stored away.

The fifth question asked is, 'How can I delete individual voice recordings?' Again, go onto the Alexa app and go to 'Settings.' Then go to 'Dialog history' to see a list of all the voice recordings. From there you can choose to listen to them or delete them.

The sixth question is, 'How do I review what I have asked my Amazon Echo?' The voice interactions with the device can be reviewed at any time. All the user needs to do is go onto the Alexa app and go to the 'Dialog history' tab that can be found in the 'Settings' section. From here, they are separated into two categories: questions and requests. They all can be either heard or deleted. Some might be marked as incomplete. This means that Alexa was not able to fully understand what was being said or asked.

The seventh question asked is, 'How does Alexa identify the wake word?' This is all seen through the software in the device. The Amazon Echo was created to be an assistant that can predict and understand your own daily routine. The keyword was created so that the technology will listen for those certain words. Since the device is powered on all the time, the device will only wait to hear 'Amazon,' 'Alexa,' or 'Simon.'

The eighth question is, 'Can the microphone on the Amazon Echo be turned off?' The simple answer to this question is 'Yes.' There are two ways to mute the microphone. One way is to use the button on the top of the device. The second way is too use the remote control. Either way, the light ring at the top of the device will and turn a solid red to show the user that it is muted.

The ninth question asked is, 'How do I make sure that my voice is streaming to the Cloud?' The best way to do this is through the wake word. Once the wake word is spoken and the device begins to respond, it can be seen that the voice is streaming through the Cloud. Otherwise, the best way is to hold the 'Talk' button on the remote and talk into the top of the remote. You know that it is working when the light ring on the top of the device turns a cyan color.

The tenth question asked is, 'Will the Amazon Echo perform better with time?' The answer is another simple 'Yes.' All the people that have developed this technology are always working to upgrade the device. Since there are many people that go to Amazon, the service team looks to have as many positive reviews as possible. With the Echo, the software constantly being ensures that it will definitely perform much better as the weeks pass by.

The final question normally asked is, 'How do I wake up Alexa?' There are three ways that this can be done. The first way is through the wake word 'Amazon,' 'Alexa,' or 'Simon.' The wake word all depends on what you set the device to recognize. The second way is to click the button on the top of the device to activate it. The third way is to hold the talk button on the remote control and speak into the top of the remote to wake up the device.

Simon Monty

Conclusion

With all the features that come with the Amazon Echo, there are many reasons to have one in your home. Whether it is for asking questions, following recipes, controlling the house, or keeping track of a schedule, this device takes the idea of a smartphone to the next level. As a basic personal assistant, the Echo offers more than most people expect. A phone can be used to look up questions online, but now you can do that without using out your phone. This is a new way to find answers. The same way that everyone replaced their flip phones with smartphones, the Amazon Echo goes beyond the smartphone in what it offers.

This book listed many benefits and many reasons to go out and purchase one of these items for yourself. It is not about wasting money. This device can play music all through your home while reminding you of your next meeting while turning a timer on for dinner. It creates a new atmosphere inside of any home, and it is always being upgraded to provide more features to the user.

Simon Monty

Thank you once again for purchasing this book! I hope that you enjoyed reading all about the benefits that the Amazon Echo has too offer.

www.ingramcontent.com/pod-product-compliance
Lightning Source LLC
Chambersburg PA
CBHW070102210526
45170CB00012B/705